THE WATCHDOG AND THE BURGLAR:
A GAME-THEORETIC PURSUIT PROBLEM

Arthur Ziffer

authorHOUSE®

AuthorHouse™
1663 Liberty Drive
Bloomington, IN 47403
www.authorhouse.com
Phone: 1-800-839-8640

Published by AuthorHouse   11/17/2014

ISBN: 978-1-4969-5434-3 (sc)
ISBN: 978-1-4969-5433-6 (e)

Library of Congress Control Number: 2014920788

CONTENTS

The Watchdog and the Burglar

by S P. Thompson and A. J. Ziffer

Naval Research Laboratory

There is presented a purposefully oversimplified game-theoretic
analysis of a competitive situation which is possibly of considerable
practical importance to certain engineering developments. For purposes
of exposition, the work will be interpreted in terms of watchdogs and
burglars. Mathematically, it is a consideration of the dependence upon
two parameters of the value of a certain matrix game played between a
watchdog and a burglar.

These parameters are interpreted as measuring the biting ability and
the stealth of the watchdog, and our principal interest is in determining
the circumstances under which a breeder of watchdogs should lessen the one
in order to increase the other, and vice versa. The game is introduced as
a mathematical device which allows the determination of the optimal behavior
in the maneuvering of the dog and the burglar for all values of the parameters.
Assuming such optimal behavior, the payoff to the breeder then becomes a function only of the parameters, making possible quantitative statements about
the relative desirability of different portions of the space of these parameters.

Hence consider the following situation. A burglar has intruded into
the watchdog's yard. The dog positions himself squarely behind the burglar
so that by making a last forward lunge he may bite the burglar with sufficient
severity to render him unfit to pursue his occupation. At this juncture, a
two-person zero-sum game is played. We suppose that the watchdog can lunge
straight ahead, bear to the left, or bear to the right; and that the burglar
is capable independently of similar maneuvers. Also let us suppose that the

burglar detects the watchdog with probability ρ. If he detects the watchdog, then he can choose any one of the three directions. If he does not detect the watchdog (which happens with probability $1-\rho$), then we assume that he goes straight ahead. In any case, both the watchdog and the burglar know that the watchdog is ignorant of whether or not the burglar has detected him. For convenience, we shall henceforth refer to the eventuality of the burglar detecting the watchdog as detection.

We express the payoff of the game as the probability of the burglar's receiving an incapacitating bite, and for simplicity assume[*] that this probability is unity if both choose identical directions and the interception is therefore squarely made, is zero if the burglar dodges left/right when the dog lunges right/left and the interception fails, and has some intermediate value λ for grazing interceptions, i.e., when one continues straight and the other deviates in either direction.

We formalize these considerations by employing the numbers 1,2,3, to indicate the directions left, straight, and right, respectively, and by employing W for watchdog and B for burglar. The following array then summarizes the foregoing payoff assignments:

W \ B	1	2	3
1	1	λ	0
2	λ	1	λ
3	0	λ	1

(1)

It is to be noted that the elements may be interpreted as payments by B to W: if the numbers representing the directions chosen differ by two, then W receives zero; if the numbers are the same, then W receives unity; and in

---

[*]At the end of the paper some generalization of these assumptions is discussed.

the case where they differ by one, W receives the intermediate reward of λ, where λ is some number such that $0 \leq \lambda \leq 1$. The parameter λ, the payoff to the watchdog for a grazing interception, can be considered directly proportional to biting ability. Finally, we remark that both W and B are aware of the values of the parameters ρ and λ in any particular play of the game.

The preceding description of the game is in extensive form. To put the game in normal form, we let w_i, $i = 1,2,3$, be the probability of W employing his i-th pure strategy, defined as:

W's i-th pure strategy: turn in direction i; and let b_j, $j = 1,2,3$, be the probability of B employing his j-th pure strategy, defined as:

B's j-th pure strategy: turn in direction j if detection occurs, otherwise take direction 2. The normal form of the game may then be represented by:

$$
\begin{array}{c|ccc}
 & b_1 & b_2 & b_3 \\
\hline
w_1 & p_{11} & p_{12} & p_{13} \\
w_2 & p_{21} & p_{22} & p_{23} \\
w_3 & p_{31} & p_{32} & p_{33} \ ,
\end{array}
\tag{2}
$$

where the p's are functions of ρ and of the elements of the array previously displayed, and where each represents the payoff which would occur if the corresponding values of w and b were unity and the remaining w's and b's were zero.

Consider p_{11}. If W lunges left, then with probability $(1-\rho)$, B fails to see W, goes straight, and the payoff is λ; while with probability ρ he sees W, goes left, and the payoff is unity. Therefore

$$p_{11} = (1-\rho)\lambda + \rho.$$

Similar reasoning shows that:

$$
\begin{aligned}
p_{12} &= p_{32} = \lambda, \\
p_{13} &= p_{31} = (1-\rho)\lambda,
\end{aligned}
\tag{3}
$$

$$p_{21} = p_{23} = \rho\lambda + (1-\rho),$$

$$p_{22} = 1.$$

The expected payoff is then:

$$\sum_{i,j=1}^{i,j=3} p_{ij} w_i b_j, \tag{4}$$

which we assume W desires to maximize by choosing the w_i's subject only to the restriction that they be non-negative and sum to unity, and which we assume B desires to minimize by choosing the b_j's subject to a similar restriction. The optimal strategies for W and B are mathematically those probability vectors $w = (w_1, w_2, w_3)$ and $b = (b_1, b_2, b_3)$ which minimax the expected payoff in the manner customary in game theory, and may be interpreted as indicating the probabilities which each in self interest should employ.

All of the optimal strategies have been obtained by examining, in the manner given in Chapter 3 of the Reference, all square sub-matrices of (2). The character of the solutions depends upon the location of the parameter point (λ, ρ) in the plane of these parameters, which plane naturally divides into three regions sketched in Fig. 1 and specified by:

$$\text{I.} \quad \rho \leq \tfrac{1}{2},$$

$$\text{II.} \quad \rho > \tfrac{1}{2}, \quad \lambda \geq \frac{2-3\rho}{2(1-2\rho)},$$

$$\text{III.} \quad \rho > \tfrac{1}{2}, \quad \lambda \leq \frac{2-3\rho}{2(1-2\rho)}.$$

The optimal strategies corresponding to each region are found to be:

In I, $w = (0,1,0)$, $b = (r,0,1-r)$, where r is any non-negative number not greater than unity. Left and right turns are superfluous pure strategies for W, as is the choice of the straight ahead direction in the case of detection for B. Hence W should always go straight, and every mixed strategy which avoids going straight in the case of detection is an optimal strategy for B.

In II, $w = (0,1,0)$, $b = (s,0,1-s)$, where s is any number such that

$$\frac{(1-2\rho)(\lambda-1)}{\rho} \le s \le \frac{\rho\lambda+(1-\rho)(1-\lambda)}{\rho}$$

The pure strategies which are superfluous in I are also superfluous in II. However, B can no longer play optimally merely by avoiding going straight in the case of detection.

In III, $w = (t, 1-2t, t)$, $b = (\frac{t}{\rho}, 1-2\frac{t}{\rho}, \frac{t}{\rho})$, where $t = \frac{1-\lambda}{3-4\lambda}$. No pure strategy is superfluous for either player in the whole region, and the optimal strategies for both are symmetric in the sense that both must play their first and third pure strategies with equal probability.

We omit from discussion the problems facing the instructors of watchdogs and burglars in teaching their pupils to randomize their jumps in accordance with optimal strategies. Assuming this to be done, we concentrate upon the problems facing the breeder of watchdogs, examining in greater detail the implications of the value of the game, defined in game theory as the expected payoff, (4), evaluated under the optimal strategies. Using V to symbolize the value, one finds that:

In I and II, $V = \rho\lambda + (1-\rho)$;

In III, $V = \frac{(1-2\lambda^2)}{(3-4\lambda)}$.

As is always the case, the value (interpreted as the probability of an incapacitating bite) is a unique function of the parameters, and inspection will show that the value is continuous across the line $\lambda = \frac{(2-3\rho)}{2(1-2\rho)}$ separating regions I and II (where V is a function both of λ and ρ) from III (where V is a function only of λ).

Fig. 2 will assist in discussing these functions and their implications for the breeder of watchdogs. The heavily drawn lines are isovalues (curves of constant V) in the (λ, ρ) plane. The isovalues for $V < \frac{1}{2}$ become of infinite slope as they cross the dotted curve separating region III from the rest of the unit square, since above this curve V is a function of λ but not of ρ. We recall that biting ability increases with λ, while stealth decreases with ρ, and that V is the probability of an incapacitating bite. As might be anticipated, V increases with λ at constant ρ. However, watchdog breeders rarely get something for nothing, and it is to be expected that in practical cases the breeder will be bound by an additional relationship connecting λ and ρ which will prohibit him from keeping ρ constant as he increases λ. Because increased biting ability in the watchdog is almost certainly obtained at the price of his easier detection by the burglar, because of the dog's larger size, it is to be expected that in practice ρ is a monotone non-decreasing function of λ. The given plot of isovalues may be used to explore the consequences of any such relationship which a practical watchdog breeder might desire to hypothesize.

Some general tendencies controlled by the form of such relationships may be easily illustrated. If the function is convex downward like the curve in Fig. 2 labelled 1, it tends to be parallel to the isovalues themselves, and the value of the game (the probability of an incapacitating bite) hence tends to vary but little as one moves along the curve. In Fig. 3, the values of the game which one obtains in this motion are plotted as a function of λ, this plot also being labelled 1. While the breeder should choose λ as large as possible in this particular case, V for $\lambda = .8$ is only about 35% larger than it is for $\lambda = 0$, and it is not obvious that the breeder should be greatly concerned about optimizing the breed.

However, the variation in V tends to be larger when ρ as a function of λ is convex upward like the curve in Fig. 2 which is labelled 2. The corresponding plot in Fig. 3 now shows that V for $\lambda = .1$ is about 65% larger than for $\lambda = .4$, where it is a minimum.

The tendency for large variation in V along the $\rho(\lambda)$ curve is even more accentuated for S-shaped curves like the one labelled 3 in Fig. 2. The corresponding curve similarly labelled in Fig. 3 shows that V at $\lambda = .1$ is now about $2\frac{1}{2}$ times greater than its minimum value at $\lambda = .25$. This circumstance would worry the authors if they were watchdog breeders, for they suspect that this is actually the type of curve to be anticipated in practice. They would expect as burglars to see small dogs with a low probability nearly independent of the size of the dog, and would expect this probability to increase very rapidly once the dog exceeded a critical size, saturating finally at a value less than unity, a value determined not by the dog, but by the burglar's imperfect alertness. (It can, of course, be true that this conjecture is correct, but that the saturation value of ρ is sufficiently small that V remains large and relatively constant over the whole of the $\rho(\lambda)$ curve, justifying less concern in selecting the value of λ than appears to be appropriate in the hypothetical example cited.)

The deep minimum in the curve labelled 3 in Fig. 3 illustrates, for the similarly-labelled S-shaped curve in Fig. 2, the danger of uncritical compromise between stealth and biting ability. Narrow-minded concentration on either of these attributes to the exclusion of the other would, in the absence of any analysis, give a watchdog of high performance, while well-intentioned but unthinking compromise is liable to unfortunate consequences.

It is interesting to consider two generalizations of this simple watchdog-burglar game. The first one involves only the following change: when it comes time for W to play, we suppose that with probability σ he receives a payoff of μ, and with probability $1-\sigma$ he plays as before. This variation is of particular interest for the case $\mu = 1$, since we may interpret the new feature as incorporating the probability σ that W perceives B's maneuver and rushes the interception squarely, and the probability $(1-\sigma)$ that W does not so observe and is forced to play as before.

The strategy payoff matrix A_σ of the new game takes the form

$$A_\sigma = (1-\sigma)A + \mu\sigma J,$$

where A is the strategy payoff matrix (2) of the original game, and J is a matrix of all ones and of the same order as A. Owing to the fact that the elements of A_σ are formed by multiplying all the elements of A by a common factor and then adding a common number to all, the optimal strategies for the A_σ game are the same as the optimal strategies for the A game, and the values of the games $V(A_\sigma)$ and $V(A)$ are related by the equation

$$V(A_\sigma) = (1-\sigma)V(A) + \mu\sigma.$$

Another generalization is a continuous analog of the original game. Let w and b be continuous variables in the interval $0 \leq w, b \leq 1$, and consider the function M formed as the sum of a function M_1 applying with probability ρ, and a function M_2 applying with probability $(1-\rho)$:

$$M(w,b) = \rho M_1(w,b) + (1-\rho) M_2(w,b),$$

where

$$M_1(w,b) \equiv (2-4\lambda)(w-b)^2 + (4\lambda-3)|w-b| + 1,$$
$$M_2(w,b) \equiv 4(\lambda-1)w^2 + 4(1-\lambda)w + \lambda.$$

The values $w,b = 0,\frac{1}{2},1$ are intended to correspond respectively to the maneuvers left, center, right in the original game. That these identifications are correct may be seen by computing the array:

$$\begin{pmatrix} M_1(0,0) & M_1(0,\frac{1}{2}) & M_1(0,1) \\ M_1(\frac{1}{2},0) & M_1(\frac{1}{2},\frac{1}{2}) & M_1(\frac{1}{2},1) \\ M_1(1,0) & M_1(1,\frac{1}{2}) & M_1(1,1) \end{pmatrix},$$

and verifying that the result is (1); and computing $M_2(0,\frac{1}{2})$, $M_2(\frac{1}{2},\frac{1}{2})$, $M_2(1,\frac{1}{2})$, and verifying that these constitute the center column of (1), that is the payoffs to W when B goes straight.

Strategies for a continuous game consist of probability distribution functions which for our game we denote W(w) and B(b). We have found optimal strategies in those regions where there were superfluous pure strategies at every parameter point, namely in regions I and II. Using the step functions $I_a(z)$ and $I_0(z)$ defined for arbitrary $0 \leq z \leq 1$ and $0 < a \leq 1$ as:

$$I_a(z) = 0, \quad \text{when } z < a,$$
$$I_a(z) = 1, \quad \text{when } z \geq a,$$
$$I_0(z) = 0, \quad \text{when } z = 0,$$
$$I_0(z) = 1, \quad \text{when } z > 0;$$

we find the following to be optimal strategies in regions I and II:

$$W(w) = I_{\frac{1}{2}}(w),$$
$$B(b) = \tfrac{1}{2}\left\{ I_0(b) + I_1(b) \right\}.$$

It will be noted that these are analogs respectively to the vectors $w = (0,1,0)$ and $b = (\tfrac{1}{2},0,\tfrac{1}{2})$, which in the original game are optimal strategies in the aforementioned regions.

In these regions the value $V(\lambda,\rho)$ in the continuous game turns out to be the same as the value in the original game. There are therefore grounds for belief that the previously discussed implications for the breeder of watchdogs stem primarily from the fundamental nature of the competitive situation treated, and not from the fact that for convenience principal attention was devoted to a discrete formalization of this situation.

Reference: J. C. C. McKinsey, "Introduction to the Theory of Games," McGraw-Hill, 1952.

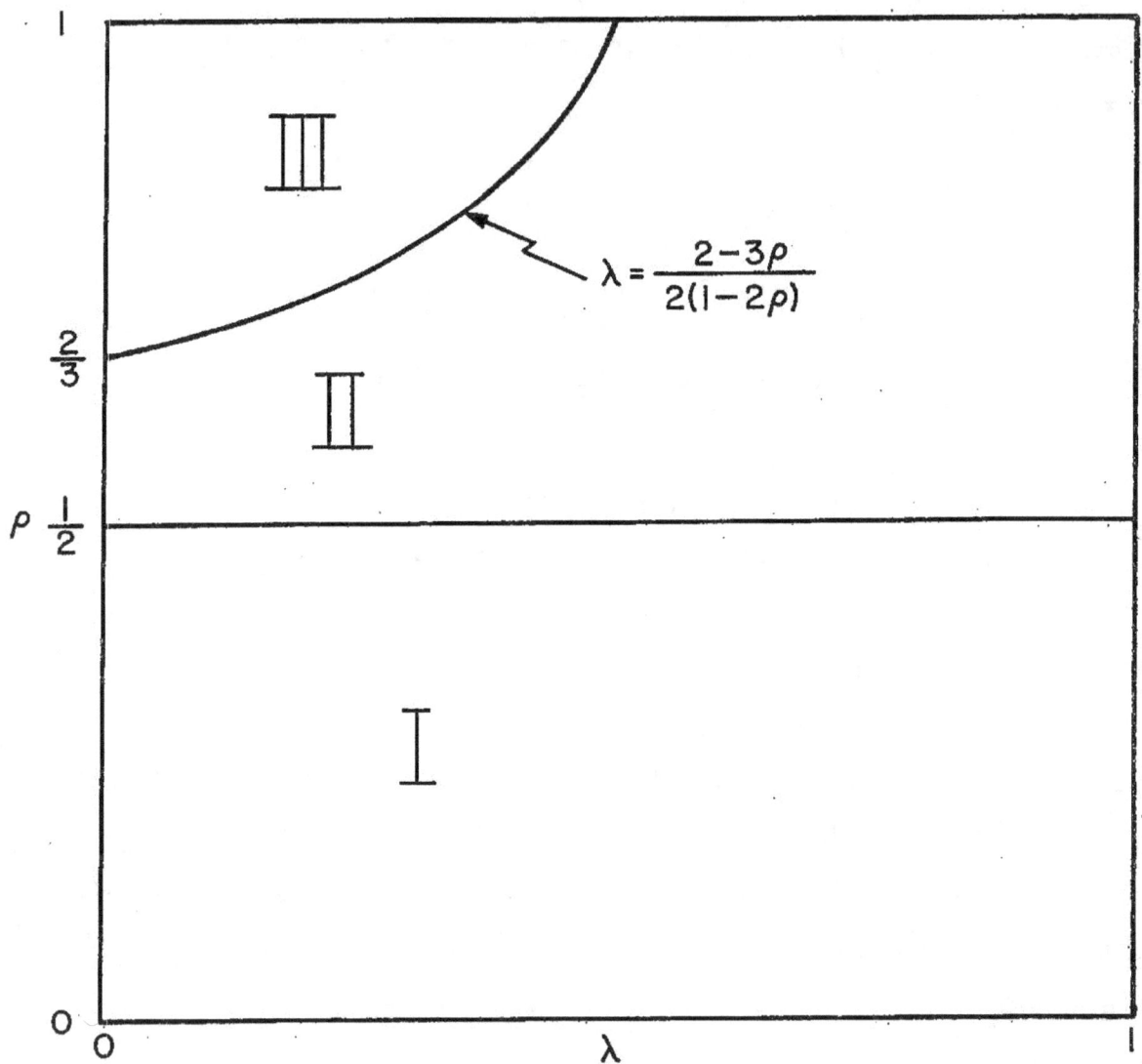

Figure 1

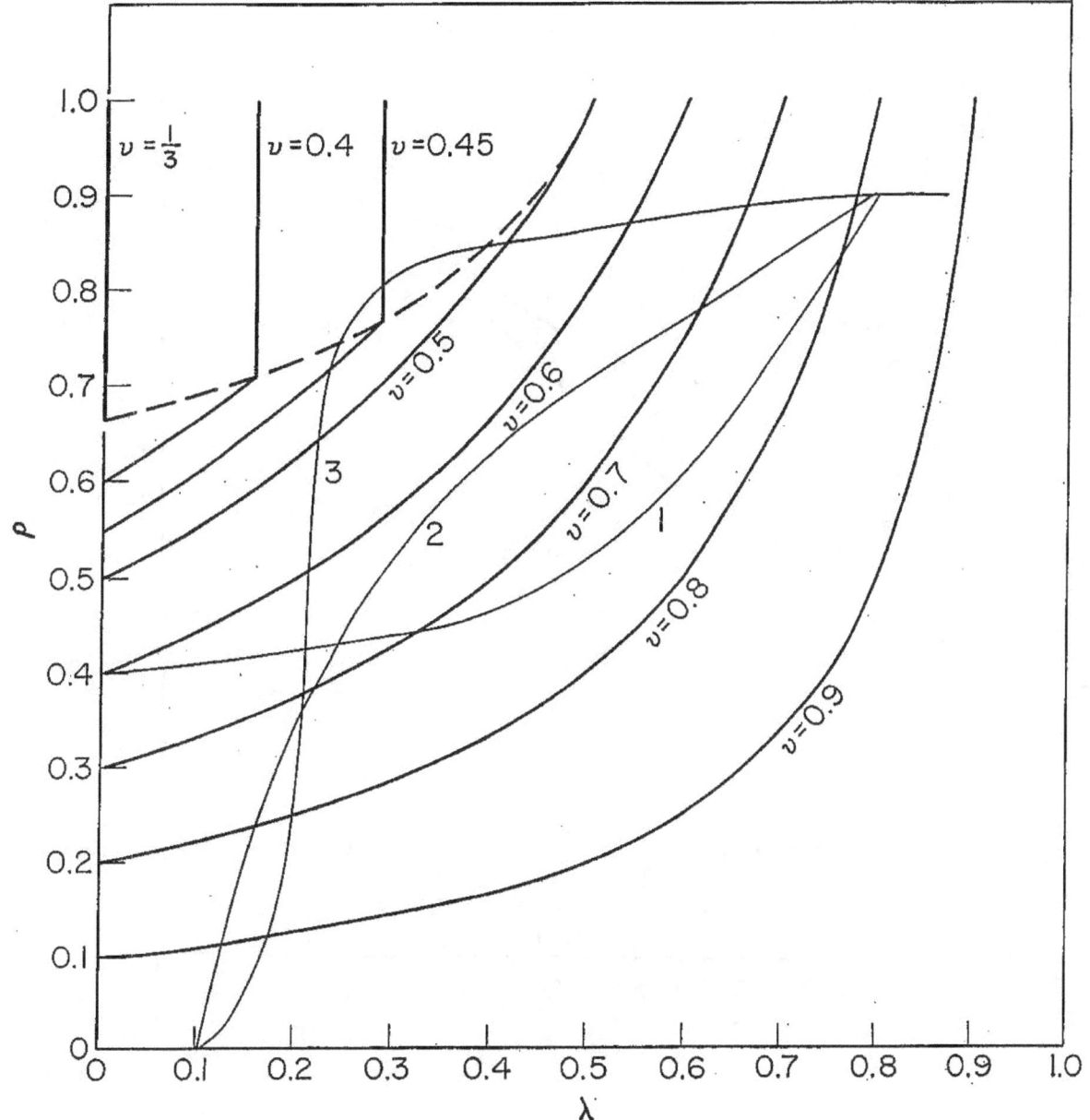

Figure 2

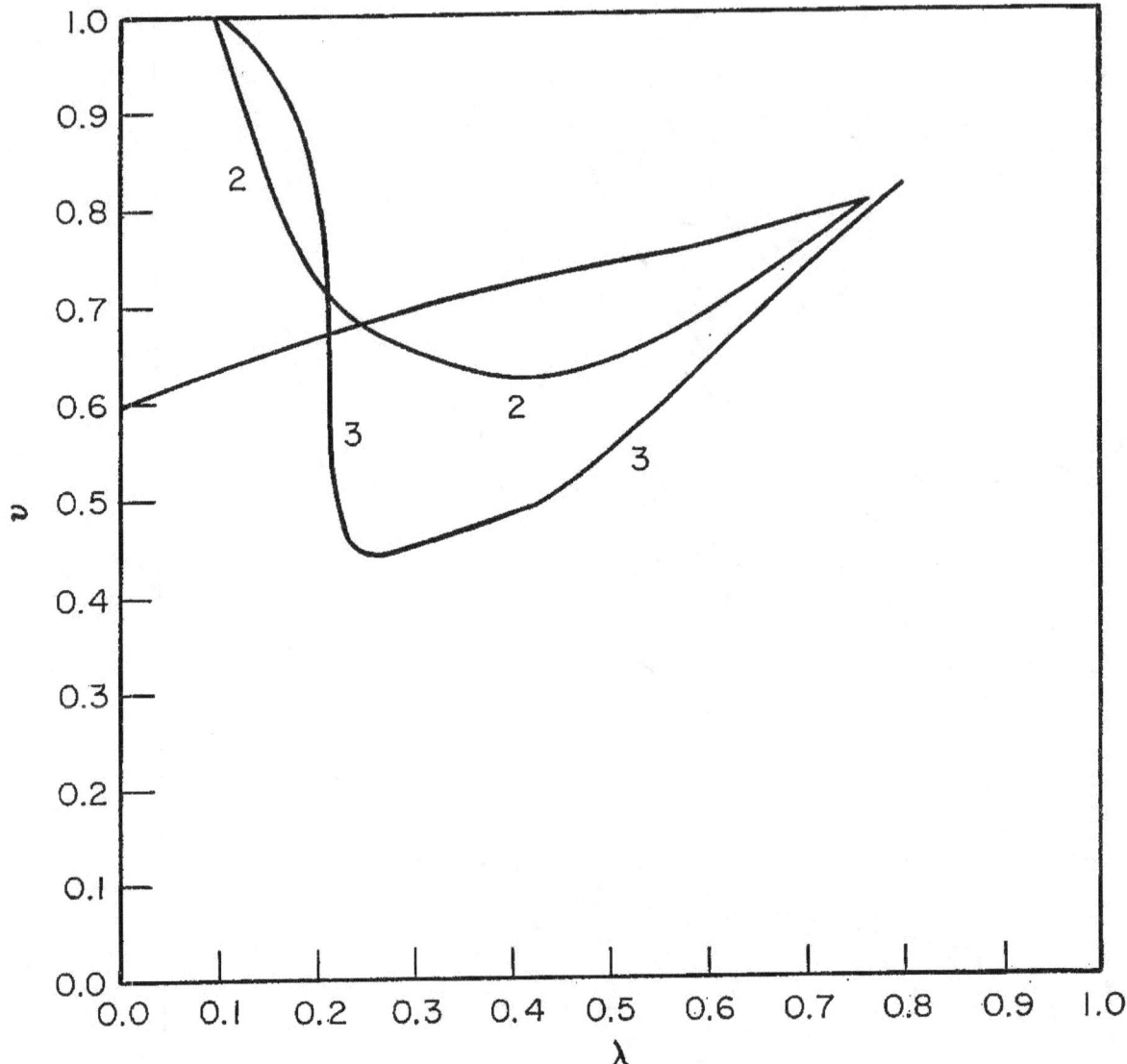

Figure 3

Mathematical Derivation of the "Watchdog and Burglar" Results

Arthur J. Ziffer

Since we will be concerned only with the mathematical derivation of the results given in the paper The Silent Watchdog and the Burglar by Sanford P. Thompson, we shall forget about the physical picture and treat it purely as a mathematical problem. As a problem in game theory, the situation reduces to the following. In move I, a chance device is used to choose a number from the set $\{1,2\}$. The chance device is so constructed that it chooses 1 with probability ρ and thus it chooses 2 with probability $1-\rho$. In move II, player B, as we shall call him, knowing what happened on the first move, chooses a number from the set $\{1,2,3\}$ if 1 was chosen on the first move and chooses 2 if 2 was chosen on the first move. In move III, the other player whom we shall call W, knowing nothing about the first two moves except that they have occurred, chooses a number from the set $\{1,2,3\}$. After these three moves, B pays W the amount 1 if they have both chosen the same number, λ if their chosen numbers differ by one, and 0 if their chosen numbers differ by two.

The graph of the above game is as follows:

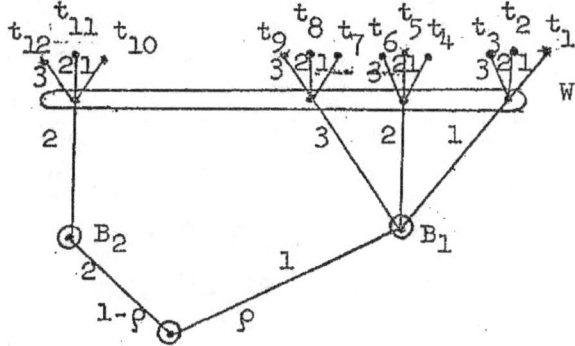

We observe from the graph that there are two information sets for B which we have labelled B_1 and B_2, and there is only one information set for W which we

2

have labelled W. The top points of the graph are labelled t_i, i = 1,2,···,12. They represent particular _plays_ of the game.

In order to solve the game we shall put it into _normal form_. To do this we must first decide just what we mean by a _strategy_ for W and a strategy for B. For player W, a strategy is simply a choice of a number from the set 1,2,3 . Let us denote W's choice of i, i = 1,2,3, as w_i, i = 1,2,3, respectively. For player B, the situation is more complicated. We recall that a strategy for a particular player is a function which is defined for each information set corresponding to the player, and whose value for each such information set is one of the alternatives there available to him (McKinsey, page 120). Hence, a single strategy for B must tell him what to do in case he is at B_1 and what to do in case he is at B_2 (letting B_1 and B_2 represent the points in the information set as well as the information set itself). Thus, B has three strategies which we shall denote as b_i, i = 1,2,3: b_i being the strategy which tells B to choose i, i = 1,2,3, respectively, in case he is at B_1 and 2 in case he is at B_2.

The next step in putting the game into normal form is to obtain _strategy payoff functions_ for B and W which we shall denote as $M_b(w_i,b_j)$ i,j = 1,2,3, and $M_w(w_i,b_j)$, i,j = 1,2,3, respectively. From the description of the game we know what the _payoffs_ to the players are, depending on what they do in a particular play of the game. That is to say, we know what the payoffs are, depending on what top point is reached at the end of a play. Or in other words, we have at our disposal _play payoff functions_ for B and W which we shall denote as $H_b(t_i)$, i = 1,2,···,12, and $H_w(t_i)$, i = 1,2,···,12, respectively. In fact,

$$H_b(t_i) = -H_w(t_i) \quad i = 1,2,···,12,$$

and

$$H_w(t_1) = 1 \qquad H_w(t_4) = \lambda \qquad H_w(t_7) = 0 \qquad H_w(t_{10}) = \lambda$$

$$H_w(t_2) = \lambda \qquad H_w(t_5) = 1 \qquad H_w(t_8) = \lambda \qquad H_w(t_{11}) = 1$$

$$H_w(t_3) = 0 \qquad H_w(t_6) = \lambda \qquad H_w(t_9) = 1 \qquad H_w(t_{12}) = \lambda.$$

Now the strategy payoff functions are supposed to tell what each player will be paid in case they use certain pure strategies. Owing to the fact that a pure strategy for B involves different choices depending on the initial chance phenomenon, we see that the strategy payoff functions are _expectations_ of each player. So if we let $p(w_i, b_j, t_k)$, $i,j = 1,2,3$, $k = 1,2,\cdots,12$, be the probability that _pure_ _strategies_ w_i, b_j result in a play ending up at t_k, we see that

$$M_b(w_i, b_j) = \sum_{k=1}^{12} H_b(t_k)\, p(w_i, b_j, t_k)$$

$$M_w(w_i, b_j) = \sum_{k=1}^{12} H_w(t_k)\, p(w_i, b_j, t_k)$$

(McKinsey, Chapter 6). Since

$$H_b(t_k) = - H_w(t_k), \qquad k = 1,2,\cdots,12,$$

we see that

$$M_b(w_i, b_j) = - M_w(w_i, b_j), \qquad i,j = 1,2,3.$$

Hence the game is _zero_ _sum_.

The necessary computations indicated above are tedious, but as an illustration we shall compute $M_w(w_1, b_1)$. Thus we must compute $p(w_1, b_1, t_k)$, $k = 1,2,\cdots,12$. $p(w_1, b_1, t_1)$, we recall, is the probability that a play will end at t_1 if B uses b_1 and W uses w_1. This is the product of three probabilities: the probability of 1 being chosen in the first move which is ρ; the probability of 1 being chosen in the second move which is 1 since B is using strategy b_1 which tells him to choose 1 if he is at B_1, and the probability of 1 being chosen in the third move which is again 1 since W is using a strategy which tells him to choose 1. Hence, $p(w_1,b_1,t_1) = (\rho)(1)(1) = \rho$. $p(w_1,b_1,t_2)$ is also the product of three probabilities: ρ, 1, and 0. The first two are obtained as in computing $p(w_1,b_1,t_1)$. The third one is 0 because if W is using strategy w_1, which tells him always to choose 1, then the probability that he will choose 2 is 0. Continuing in this way, we get

$$p(w_1,b_1,t_1) = (\rho)(1)(1) = \rho$$
$$p(w_1,b_1,t_2) = (\rho)(1)(0) = 0$$
$$p(w_1,b_1,t_3) = (\rho)(1)(0) = 0$$
$$p(w_1,b_1,t_4) = (\rho)(0)(1) = 0$$
$$p(w_1,b_1,t_5) = (\rho)(0)(0) = 0$$
$$p(w_1,b_1,t_6) = (\rho)(0)(0) = 0$$
$$p(w_1,b_1,t_7) = (\rho)(0)(1) = 0$$
$$p(w_1,b_1,t_8) = (\rho)(0)(0) = 0$$
$$p(w_1,b_1,t_9) = (\rho)(0)(0) = 0$$
$$p(w_1,b_1,t_{10}) = (1-\rho)(1)(1) = 1-\rho$$
$$p(w_1,b_1,t_{11}) = (1-\rho)(1)(0) = 0$$
$$p(w_1,b_1,t_{12}) = (1-\rho)(1)(0) = 0;$$

and thus

$$\begin{aligned}
M_w(w_1,b_1) &= (1)(\rho) + (\lambda)(0) + (0)(0)\\
&\quad + (\lambda)(0) + (1)(0) + (\lambda)(0)\\
&\quad + (0)(0) + (\lambda)(0) + (1)(0)\\
&\quad + (\lambda)(1-\rho) + (1)(0) + (1)(0)\\
&= \rho + (1-\rho)\lambda .
\end{aligned}$$

Now we have put the game into normal form. That is to say, we have what can be considered as pure strategies, and we know what the payoffs are for these pure strategies. By letting

$$a_{ij} = M_w(w_i,b_j), \quad i,j = 1,2,3$$

we are able to represent this game in matrix form:

$$\begin{pmatrix} \rho+(1-\rho)\lambda & \lambda & (1-\rho)\lambda \\ \rho\lambda+(1-\rho) & 1 & \rho\lambda+(1-\rho) \\ (1-\rho)\lambda & \lambda & \rho+(1-\rho)\lambda \end{pmatrix}$$

The i-th row gives the payoffs that W will receive if he uses strategy w_i; the j-th column gives the payoffs that B will have to make if he uses strategy b_j.

Let us now solve this game. Naturally, we first look for saddle point solutions. In all that follows, we restrict ourselves to values of λ such that $0 \le \lambda \le 1$ and of course ρ, being a probability, is such that $0 \le \rho \le 1$. We shall show that a_{21} and a_{23} are saddle points for ρ, $0 \le \rho \le \frac{1}{2}$ and for λ, $0 \le \lambda \le 1$. This means that for those particular values of ρ and λ the following inequalities must be true

$$a_{21} - a_{11} \ge 0$$
$$a_{21} - a_{31} \ge 0$$
$$a_{23} - a_{13} \ge 0$$
$$a_{23} - a_{33} \ge 0$$
$$a_{22} - a_{21} \ge 0$$
$$a_{22} - a_{23} \ge 0.$$

Since $a_{11} = a_{33}$, $a_{13} = a_{31}$ and $a_{21} = a_{23}$ (which must be the case if a_{21} and a_{23} are both to be saddle points), our problem is reduced to showing that for $\rho \le \frac{1}{2}$ the following must be true

$$a_{21} - a_{11} \ge 0$$
$$a_{21} - a_{31} \ge 0$$
$$a_{22} - a_{21} \ge 0.$$

The truth of the first is obtained by noticing that

$$a_{21} - a_{11} = [\rho\lambda+(1-\rho)] - [\rho+(1-\rho)\lambda]$$
$$= \rho(\lambda-1) + (1-\rho)(1-\lambda)$$
$$= (1-\lambda)(1-2\rho)$$
$$\geq 0 \quad \text{if} \quad \rho \leq \tfrac{1}{2}.$$

The second from

$$a_{21} - a_{31} = [\rho\lambda+(1-\rho)] - [(1-\rho)\lambda]$$
$$= \rho\lambda + (1-\rho)(1-\lambda)$$
$$\geq 0.$$

The third

$$a_{22} - a_{21} = 1 - [\rho\lambda+(1-\rho)]$$
$$= \rho - \rho\lambda$$
$$= \rho(1-\lambda)$$
$$\geq 0.$$

Hence a_{21} and a_{23} are column maxima and row minima, or saddle points for ρ, $0 \leq \rho \leq \tfrac{1}{2}$ and λ, $0 \leq \lambda \leq 1$. Since both saddle points are in the second row, $(0,1,0)$ is an optimal solution for W; and since one of the saddle points is in the first column and one is in the third, $(k,0,1-k)$ where k is any number such that $0 \leq k \leq 1$ are optimal solutions for B. The value of the game is of course the saddle value itself

$$\rho\lambda+(1-\rho).$$

Further examination of A will show that there are no other saddle pints.

Now let us look for solutions which are not saddle point solutions. We use the method given in Chapter 3 of McKinsey. In the following we shall use adj A to mean the adjoint of matrix A, A^T to mean the transpose of matrix A, $\vec{X}^T$ to mean the column vector formed from the row vector $\vec{X}$, and J will be the symbol of a vector whose components are all ones, $J = (1,1,\cdots,1)$, the number of ones to be determined by what is necessary to allow one to perform

the particular matrix or vector multiplication at hand. We now summarize
the method. Consider all square submatrices of the matrix which represents
the game. Apply a certain process, which will be described later, to these
submatrices. Sometimes this process yields a solution. Consider the set of
all solution vectors for the row player. Form all possible convex linear
combinations. We now have, according to the theorem in Chapter 3 of McKinsey,
all solution vectors for the row player. The same process gives all solutions
for the column player. We note that saddle point solutions will be given by
one by one submatrices as will be seen when we describe the process. Hence
if all saddle point solutions can be found by inspection, it is not necessary
to examine one by one submatrices. Finally, let us describe the process by
which a square submatrix B of matrix A can give rise to a solution.

1) Compute

$$V = \frac{|B|}{J(\text{adj}B)J^T}$$

$$\vec{X} = \frac{J\,\text{adj}B}{J(\text{adj}B)J^T} \qquad \vec{Y} = \frac{J(\text{adj}B)^T}{J(\text{adj}B)\,J^T} .$$

2) Supposing $\vec{X} = (x_1, \cdots, x_n)$ and $\vec{Y} = (y_1, \cdots, y_n)$, see if the
following are true:

   a) $0 \le x_i \le 1, \quad i = 1, \cdots, n$

   $0 \le y_i \le 1, \quad i = 1, \cdots, n;$

   b) $\displaystyle\sum_{i=1}^{n} x_i = 1$

   $\displaystyle\sum_{i=1}^{n} y_i = 1.$

If a) and b) are not both satisfied, then discard B. If a) and
b) are satisfied, then

3) Form the extended vectors $\vec{X}_E$ and $\vec{Y}_E$ out of $\vec{X}$ and $\vec{Y}$ by putting
in zeros exactly where rows and columns were deleted from the
original matrix A to get the submatrix B.

4) Compute $\vec{X}_E A$ and $A\vec{Y}_E^T$. If both the following are true

$$\vec{X}_E A \geq VJ$$

$$A\vec{Y}_E^T \leq VJ^T$$

then $\vec{X}_E$ and $\vec{Y}_E$ is a solution of the game and V is the

value. If such is not the case, then discard B.

Applying this method to $A = (a_{ij})$, $i,j = 1,2,3$, we find that only A

itself and the submatrices

$$\begin{pmatrix} a_{11} & a_{13} \\ a_{21} & a_{23} \end{pmatrix} \quad \text{and} \quad \begin{pmatrix} a_{21} & a_{23} \\ a_{31} & a_{33} \end{pmatrix}$$

give rise to solutions. For example: Let

$$B = \begin{pmatrix} a_{11} & a_{13} \\ a_{21} & a_{23} \end{pmatrix} = \begin{pmatrix} \rho+(1-\rho)\lambda & (1-\rho)\lambda \\ \rho\lambda+(1-\rho) & \rho\lambda+(1-\rho) \end{pmatrix}$$

1) $|B| = \rho^2\lambda + \rho(1-\rho)$

$$\text{adj } B = \begin{pmatrix} \rho\lambda+(1-\rho) & -(1-\rho)\lambda \\ -\rho\lambda-(1-\rho) & \rho+(1-\rho)\lambda \end{pmatrix}$$

$$J \text{ adj } B = (0,\rho)$$

$$J(\text{adj } B)^T = \left(\rho\lambda+(1-\rho)(1-\lambda), (2\rho-1)(1-\lambda)\right)$$

$$J(\text{adj } B)J^T = \rho.$$

Hence

$$V = \rho\lambda+(1-\rho)$$

$$\vec{X} = (0,1) \qquad \vec{Y} = \left(\frac{\rho\lambda+(1-\rho)(1-\lambda)}{\rho}, \frac{(2\rho-1)(1-\lambda)}{\rho}\right)$$

2) $\vec{X} = (x_1,x_2) = (0,1)$ obviously satisfies

$$0 \leq x_1 \leq 1$$

$$0 \leq x_2 \leq 1$$

$$x_1 + x_2 = 1$$

$\vec{Y} = (y_1,y_2)$ satisfies analogous conditions if $\rho \geq \frac{1}{2}$. Obviously y_1 and

y_2 are nonnegative for these values of ρ. Hence it remains to show that for these values of ρ

$$y_1 \leq 1$$

$$y_2 \leq 1$$

$$y_1 + y_2 \leq 1.$$

The first inequality follows from the following chain of implications:

$$\rho \geq \tfrac{1}{2}$$

$$\Rightarrow 1-2\rho \leq 0$$

$$\Rightarrow (1-2\rho)(1-\lambda) \leq 0$$

$$\Rightarrow (1-\rho)(1-\lambda) - \rho(1-\lambda) \leq 0$$

$$\Rightarrow (1-\rho)(1-\lambda) + \rho\lambda \leq \rho$$

$$\Rightarrow \frac{(1-\rho)(1-\lambda)+\rho\lambda}{\rho} \leq 1$$

and hence

$$y_1 = \frac{(1-\rho)(1-\lambda)+\rho\lambda}{\rho} \leq 1.$$

The second inequality follows from

$$\tfrac{1}{2} \leq \rho \leq 1$$

$$\Rightarrow 1 \leq \frac{1}{\rho} \leq 2$$

$$\Rightarrow -1 \geq -\frac{1}{\rho} \geq -2$$

$$\Rightarrow 1 \geq 2 - \frac{1}{\rho}$$

$$\Rightarrow 1 \geq (2-\frac{1}{\rho})(1-\lambda)$$

$$\Rightarrow 1 \geq \frac{(2\rho-1)(1-\lambda)}{\rho}$$

and hence

$$y_2 = \frac{(2\rho-1)(1-\lambda)}{\rho} \leq 1.$$

The third relation follows from a simple computation.

We remark that we used a little hindsight in obtaining the above restriction on $\rho(\rho \geq \tfrac{1}{2})$.

For convenience in later computations, we restrict ρ such that $\rho > \tfrac{1}{2}$. The solution that arises from $\rho = \tfrac{1}{2}$ is easily seen to be one of our saddle point solutions so we are not really losing anything.

3) Since the submatrix B was obtained from A by deleting the third row and second column,

$$\vec{X}_E = (0,1,0) \quad \text{and} \quad \vec{Y}_E = \left(\frac{\rho\lambda+(1-\rho)(1-\lambda)}{\rho}, 0, \frac{(2\rho-1)(1-\lambda)}{\rho}\right)$$

4) Now we must show that

$$\vec{X}_E A \geq VJ = (V,V,V)$$

and

$$A\vec{Y}_E^T \leq V \cdot J^T = \begin{pmatrix} V \\ V \\ V \end{pmatrix}.$$

Since

$$\vec{X}_E A = (\rho\lambda+(1-\rho), 1, \rho\lambda+(1-\rho)),$$
$$V = \rho\lambda+(1-\rho),$$

and

$$1 \geq 1-\rho(1-\lambda) = \rho\lambda+(1-\rho) = V,$$

we have the first inequality. To obtain the second, we let

$$\vec{Y}_E = (y_1, 0, y_3)$$

and our problem is now to show that

$$a_{11}\,y_1 + a_{13}\,y_3 \leq V$$
$$a_{21}\,y_1 + a_{23}\,y_3 \leq V$$
$$a_{31}\,y_1 + a_{33}\,y_3 \leq V.$$

The second relation follows from the fact that

$$a_{21} = a_{23} = \rho\lambda+(1-\rho) = V$$
$$y_1 + y_3 = 1,$$

and hence

$$a_{21}\,y_1 + a_{23}\,y_3 = V(y_1+y_3) = V.$$

The third relation follows from

$$a_{11} = \rho+(1-\rho)\lambda \qquad a_{13} = (1-\rho)\lambda$$
$$a_{11}\,y_1 + a_{13}\,y_3 = \rho\,y_1 + (1-\rho)\lambda\,(y_1+y_3)$$
$$= \rho\lambda+(1-\rho) = V.$$

And now back to the first

$$a_{31}y_1 + a_{33}y_3 = (1-\rho)\lambda(y_1+y_3) + \rho\, y_3$$
$$= (1-\rho)\lambda + (2\rho-1)(1-\lambda)$$
$$= \lambda - \rho\lambda + 2\rho - 2\rho\lambda - 1 + \lambda$$
$$= [\rho\lambda + (1-\rho)] - [4\rho\lambda - 3\rho + 2 - 2\lambda]$$
$$= V - [-2\lambda(1-2\rho) + (2-3\rho)]$$

which is $\leq V$ iff

$$-2\lambda(1-2\rho) + (2-3\rho) \geq 0$$

or, since $\rho > \tfrac{1}{2}$ (which is why we discarded $\rho = \tfrac{1}{2}$ before), iff

$$\lambda \geq \frac{2-3\rho}{2(1-2\rho)}$$

which is another restriction on the region of validity of the solution. Hence

$$V = \rho\lambda + (1-\rho),$$
$$\vec{X}_E = (0,1,0) \quad \text{and} \quad \vec{Y}_E = \left(\frac{\rho\lambda+(1-\rho)(1-\lambda)}{\rho}, 0, \frac{(2\rho-1)(1-\lambda)}{\rho}\right)$$

is the value and a solution of the game for $\rho > \tfrac{1}{2}$ and $\lambda \geq \dfrac{2-3\rho}{2(1-2\rho)}$.

For the same values of ρ and λ as above, the submatrix

$$\begin{pmatrix} a_{21} & a_{23} \\ a_{31} & a_{33} \end{pmatrix}$$

gives the solution

$$\vec{X}_E = (0,1,0) \qquad \vec{Y}_E = \left(\frac{(2\rho-1)(1-\lambda)}{\rho}, 0, \frac{\rho\lambda+(1-\rho)(1-\lambda)}{\rho}\right)$$

The value, of course, is the same since we are in the same region of parameter space and the value of a game is always unique.

The 3×3 matrix A itself gives a solution for $\rho > \tfrac{1}{2}$ and $\lambda \leq \dfrac{2-3\rho}{2(1-2\rho)}$ which is

$$\vec{X}_E = (z, 1-2z, z) \qquad \vec{Y}_E = \left(\frac{z}{\rho}, 1-2\frac{z}{\rho}, \frac{z}{\rho}\right)$$

where

$$z = \frac{1-\lambda}{3-4\lambda}.$$

The value of the game in this case is

$$V = \frac{1-2\lambda^2}{3-4\lambda}.$$

It does not have to be the same as the previous value since we are now in a different region of the parameter space.

Now let us summarize. We use a graph to help us, and we return to our previous notation of using W for the row player and B for the column player.

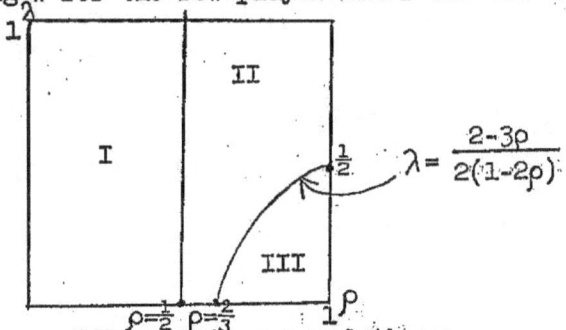

Region I ($\rho \leq \frac{1}{2}$) is our saddle point area and there

$$V = \rho\lambda + (1-\rho)$$

and

$$W = (0,1,0) \qquad B = (k,0,1-k)$$

where k is any number such that $0 \leq k \leq 1$.

Region II ($\rho > \frac{1}{2}$, $\lambda \geq \frac{2-3\rho}{2(1-2\rho)}$) is the region where we have two distinct solutions. Hence any convex linear combination of them is also a solution. And since the set of all convex linear combinations of two points is the line connecting them we have

$$V = \rho\lambda + (1-\rho)$$

and

$$W = (0,1,0) \qquad B = (u,0,v)$$

where u and v are any numbers such that

$$\frac{(1-2\rho)(\lambda-1)}{\rho} \leq u \leq \frac{\rho\lambda+(1-\rho)(1-\lambda)}{\rho}$$

$$\frac{(1-2\rho)(\lambda-1)}{\rho} \leq v \leq \frac{\rho\lambda+(1-\rho)(1-\lambda)}{\rho}$$

Region III ($\rho > \frac{1}{2}$, $\lambda \leq \frac{2-3\rho}{2(1-2\rho)}$) possesses only one solution and we have

$$V = \frac{1-2\lambda^2}{3-4\lambda}$$

and

$$W = (z, 1-2z, z) \qquad B = (\frac{z}{\rho}, 1-2\frac{z}{\rho}, \frac{z}{\rho})$$

where

$$z = \frac{1-\lambda}{3-4\lambda} .$$

We notice that the solution in region III has the property that the first and third components of both solution vectors are equal. In regions I and II, the only solution with this property is

$$W = (0,1,0) \qquad B = (\tfrac{1}{2},0,\tfrac{1}{2}).$$

These are exactly the results used in the paper The Silent Watchdog and the Burglar. The last statement, of course, depends upon the fact that the set of points of the ρ,λ square

$$\left\{ (\rho,\lambda) \mid \rho > \tfrac{1}{2}, \ \lambda \leq \frac{2-3\rho}{2(1-2\rho)} \right\}$$

is the same as the set

$$\left\{ (\rho,\lambda) \mid \lambda \leq \tfrac{1}{2}, \ \rho \geq \frac{2(1-\lambda)}{3-4\lambda} \right\}$$

This follows by observing that

$$\rho > \tfrac{1}{2}, \ \lambda \leq \frac{2-3\rho}{2(1-2\rho)} \Rightarrow 2(1-2\rho)\lambda \geq 2-3\rho$$

$$\Rightarrow \rho(3-4\lambda) \geq 2(1-\lambda)$$

$$\Rightarrow \rho \geq \frac{2(1-\lambda)}{3-4\lambda} \text{ if } 3-4\lambda > 0;$$

and $3-4\lambda > 0$ since

$$\rho > \tfrac{1}{2}, \quad \lambda \leq \frac{2-3\rho}{2(1-2\rho)}$$

$$\Rightarrow \lambda \leq \tfrac{1}{2}.$$

and conversely

An Extension of the "Watchdog and
Burglar" Problem
Arthur J. Ziffer

This report refers to the paper The Silent Watchdog and the Burglar
by Sanford P. Thompson and the report Mathematical Derivation of the "Watchdog and Burglar" Results by the present author. The generalization to be
given here results from a straight application of the methods presented in
the last mentioned report.

Hence consider the following game: In move I, a chance device is used
to choose a letter from the set $\{\alpha, \beta\}$. It chooses α with probability ρ and
thus it chooses β with probability $1-\rho$. In move II, player B chooses a
number from the set $\{1,2,3\}$ if α was chosen in move I, and he chooses 2 if
β was chosen in move I. In move III, a chance device is used to choose a
letter from the set $\{\gamma, \delta\}$. It chooses γ with probability σ and thus it
chooses δ with probability $1-\sigma$. In move IV, player W chooses ϵ if γ was
chosen in move III, and he chooses a number from the set $\{1,2,3\}$ if δ was
chosen in move III. At the end of all this, B then pays W μ if W chose ϵ,
1 if they both chose the same number, λ if their chosen numbers differ by 1,
and 0 if their chosen numbers differ by 2.

Representing this game by a graph, we get Fig. 1. t_i, $i = 1,2,\cdots,16$
are the top points. B_1 and B_2 are information sets for player B, and W_1 and
W_2 are information sets for player W.

Let $H_b(t_i)$ and $H_w(t_i)$, $i = 1,2,\cdots,16$ be the play payoff functions for
W and B respectively; then

$$H_b(t_i) = - H_w(t_i), \quad i = 1,2,\cdots,16$$

and

$$H_w(t_1) = \mu \qquad H_w(t_5) = \mu \qquad H_w(t_9) = \mu \qquad H_w(t_{13}) = \mu$$
$$H_w(t_2) = 1 \qquad H_w(t_6) = \lambda \qquad H_w(t_{10}) = 0 \qquad H_w(t_{14}) = \lambda$$
$$H_w(t_3) = \lambda \qquad H_w(t_7) = 1 \qquad H_w(t_{11}) = \lambda \qquad H_w(t_{15}) = 1$$
$$H_w(t_4) = 0 \qquad H_w(t_8) = \lambda \qquad H_w(t_{12}) = 1 \qquad H_w(t_{16}) = \lambda \ .$$

Let b_i, $i = 1,2,3$, be the pure strategies for B which tell him to choose i, $i = 1,2,3$, respectively, in move II if α was chosen in move I and to choose 2 if β was chosen in move I.

Let w_i, $i = 1,2,3$, be the pure strategies for W which tell him to choose i, $i = 1,2,3$, respectively, in move IV if δ was chosen in move III and to choose ϵ if δ was chosen in move III.

Let $p(w_i, b_j, t_k)$, $i, j = 1,2,3$, $k = 1,2,\cdots,16$ be the probability that a play will end up at top point t_k if B uses strategy b_i and W uses strategy w_j. As an illustration, let us compute $p(w_1, b_1, t_k)$, $k = 1,2,\cdots,16$. Elementary probability theory tells us that $p(w_1, b_1, t_1)$, for example, is the product of the following four probabilities: the probability that α is chosen in move I which is ρ; the probability that 1 is chosen in move II which is 1 since B is using strategy b_1 which tells him to choose 1 in the event α was chosen in move I; the probability that δ is chosen in move III which is σ, and the probability that ϵ is chosen in move IV which is 1 since W is using strategy w_1 which tells him to choose ϵ in the event that δ was chosen in move III. Hence

$$p(w_1, b_1, t_1) = (\rho)(1)(\sigma)(1) = \rho\sigma.$$

Continuing in this way, we get

$$p(w_1, b_1, t_1) = (\rho)(1)(\sigma)(1) = \rho\sigma$$
$$p(w_1, b_1, t_2) = (\rho)(1)(1-\sigma)(1) = \rho(1-\sigma)$$
$$p(w_1, b_1, t_3) = (\rho)(1)(1-\sigma)(0) = 0$$
$$p(w_1, b_1, t_4) = (\rho)(1)(1-\sigma)(0) = 0$$

$$p(w_1, b_1, t_5) = (\rho)(0)(\sigma)(1) = 0$$

$$p(w_1, b_1, t_6) = (\rho)(0)(1-\sigma)(1) = 0$$

$$p(w_1, b_1, t_7) = (\rho)(0)(1-\sigma)(0) = 0$$

$$p(w_1, b_1, t_8) = (\rho)(0)(1-\sigma)(0) = 0$$

$$p(w_1, b_1, t_9) = (\rho)(0)(\sigma)(1) = 0$$

$$p(w_1, b_1, t_{10}) = (\rho)(0)(1-\sigma)(1) = 0$$

$$p(w_1, b_1, t_{11}) = (\rho)(0)(1-\sigma)(0) = 0$$

$$p(w_1, b_1, t_{12}) = (\rho)(0)(1-\sigma)(0) = 0$$

$$p(w_1, b_1, t_{13}) = (1-\rho)(1)(\sigma)(1) = (1-\rho)\sigma$$

$$p(w_1, b_1, t_{14}) = (1-\rho)(1)(1-\sigma)(1) = (1-\rho)(1-\sigma)$$

$$p(w_1, b_1, t_{15}) = (1-\rho)(1)(1-\sigma)(0) = 0$$

$$p(w_1, b_1, t_{16}) = (1-\rho)(1)(1-\sigma)(0) = 0 .$$

Finally, let $M_b(w_i, b_j)$ and $M_w(w_i, b_j)$, $i,j = 1,2,3$, be the strategy payoff functions,

$$M_b(w_i, b_j) = \sum_{k=1}^{16} H_b(t_k)\, p(w_i, b_j, t_k)$$

$$M_w(w_i, b_j) = \sum_{k=1}^{16} H_w(t_k)\, p(w_i, b_j, t_k)$$

We note that

$$M_b(w_i, b_j) = - M_w(w_i, b_j)$$

which means that the game is zero sum. As an illustration, let us compute

$$M_w(w_1, b_1) = \sum_{k=1}^{16} H_w(t_k)\, p(w_1, b_1, t_k)$$

$$= [\mu][\rho\sigma] + [1][\rho(1-\sigma)] + [\lambda][0] + [0][0]$$

$$+ [\mu][0] + [\lambda][0] + [1][0] + [\lambda][0]$$

$$+ [\mu][0] + [0][0] + [\lambda][0] + [1][0]$$

$$+ [\mu][(1-\rho)\sigma] + [\lambda][(1-\rho)(1-\sigma)] + [1][0] + [\lambda][0]$$

$$= \mu\rho\sigma + \rho(1-\sigma) + \mu(1-\rho)\sigma + \lambda(1-\rho)(1-\sigma)$$

$$= (1-\sigma)[\rho + \lambda(1-\rho)] + \mu\sigma .$$

Continuing in this way, we get, letting $a_{ij} = M_W(w_i, b_j)$ and $A_\sigma = (a_{ij})$,

$$A_\sigma = \begin{pmatrix} (1-\sigma)[\rho+(1-\rho)\lambda]+\mu\sigma & (1-\sigma)[\lambda]+\mu\sigma & (1-\sigma)[(1-\rho)\lambda]+\mu\sigma \\ (1-\sigma)[\rho\lambda+(1-\rho)]+\mu\sigma & (1-\sigma)[1]+\mu\sigma & (1-\sigma)[\rho\lambda+(1-\rho)]+\mu\sigma \\ (1-\sigma)[(1-\rho)\lambda]+\mu\sigma & (1-\sigma)[\lambda]+\mu\sigma & (1-\sigma)[\rho+(1-\rho)\lambda]+\mu\sigma \end{pmatrix}$$

which is the same as

$$(1-\sigma)\begin{pmatrix} \rho+(1-\rho)\lambda & \lambda & (1-\rho)\lambda \\ \rho\lambda+(1-\rho) & 1 & \rho\lambda+(1-\rho) \\ (1-\rho)\lambda & \lambda & \rho+(1-\rho)\lambda \end{pmatrix} + \mu\sigma\begin{pmatrix} 1 & 1 & 1 \\ 1 & 1 & 1 \\ 1 & 1 & 1 \end{pmatrix}$$

which we write as

$$(1-\sigma)A + \mu\sigma J$$

with obvious definitions for A and J.

The matrix A_σ is the matrix representation of the game in normal form. From the identity

$$A_\sigma = (1-\sigma)A + \mu\sigma J$$

and some theorems from game theory, we know that the set of solutions for the game represented by A_σ is the same as the set of solutions for the game represented by A and that the values of the games satisfy the following equation

$$v(A_\sigma) = (1-\sigma)v(A) + \mu\sigma$$

where $v(A_\sigma)$ and $v(A)$ are respectively the values of the games represented by the matrices A_σ and A (see page 46 of "The Compleat Strategyst" by J. D. Williams).

Hence, using the results of the report Mathematical Derivation of "Watchdog and Burglar" Results, where we solved a game which had matrix representation A, we find that:

for $\rho \leq \frac{1}{2}$,

$$v = (1-\sigma)[\rho\lambda+(1-\rho)] + \mu\sigma$$
$$W = (0,1,0) \qquad B = (k,0,1-k)$$

where k is any value such that $0 \leq k \leq 1$;

for $\rho > \frac{1}{2}$, $\lambda \geq \dfrac{2-3\rho}{2(1-2\rho)}$,

$$v = (1-\sigma)\big[\rho\lambda+(1-\rho)\big] + \mu\sigma$$

$$W = (0,1,0) \qquad B = (r,0,s)$$

where r and s are any numbers such that

$$\frac{(1-2\rho)(\lambda-1)}{\rho} \leq r \leq \frac{\rho\lambda+(1-\rho)(1-\lambda)}{\rho}$$

$$\frac{(1-2\rho)(\lambda-1)}{\rho} \leq s \leq \frac{\rho\lambda+(1-\rho)(1-\lambda)}{\rho};$$

and for $\rho > \frac{1}{2}$, $\lambda \leq \dfrac{2-3\rho}{2(1-2\rho)}$,

$$v = (1-\sigma)\left[\frac{1-2\lambda^2}{3-4\lambda}\right] + \mu\sigma$$

$$W = (z, 1-2z, z) \qquad B = \left(\frac{z}{\rho}, 1-2\frac{z}{\rho}, \frac{z}{\rho}\right)$$

where $z = \dfrac{1-\lambda}{3-4\lambda}$.

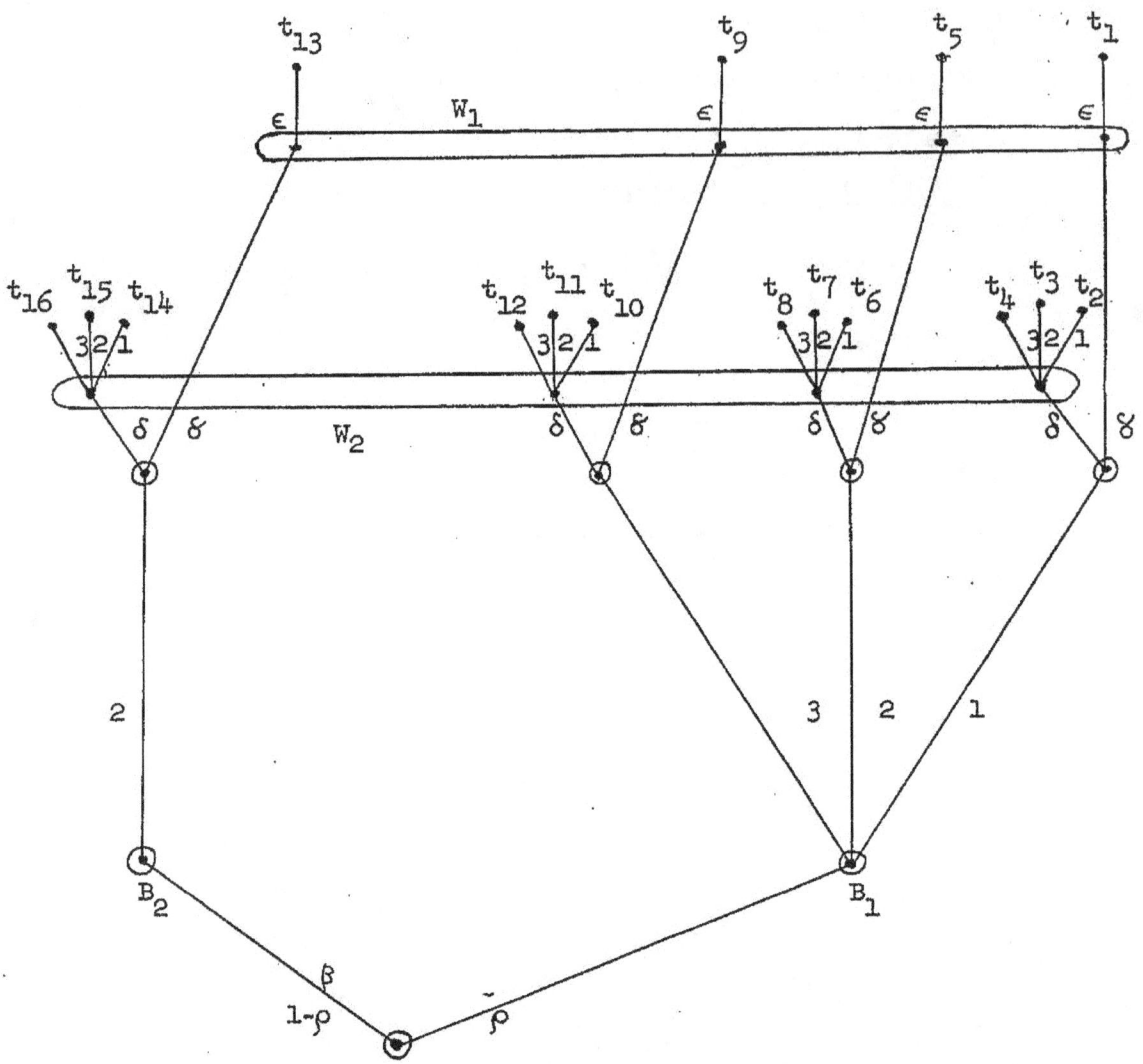

Figure 1

A Recursive Form of the Watchdog
and Burglar Problem
Arthur J. Ziffer

We shall consider the following simplified form of the problem. At move I, player B chooses a number b from the set $\{1,2,3\}$. At move II, player W, not knowing what B chose in move I, also chooses a number w from the set $\{1,2,3\}$. Let us let (b,w) represent B's choice of b in move I and W's choice of w in move II. Then in the case that

$$(b,w) \in \{(1,1),(2,2),(3,3)\}$$

B pays W the value one and the play is over. In the case

$$(b,w) \in \{(1,2),(2,1),(2,3),(3,2)\}$$

then B pays W the value λ where λ is a number such that $0 \leq \lambda \leq 1$ and the play is over. Finally, in the case

$$(b,w) \in \{(1,3),(3,1)\}$$

the players start from move I again. This continues until one of the two previous stops are reached.

This is a recursive game. Let us designate it as Γ. Then it can be characterized by the matrix

$$\Gamma : \begin{pmatrix} 1 & \lambda & \Gamma \\ \lambda & 1 & \lambda \\ \Gamma & \lambda & 1 \end{pmatrix}$$

The basic paper on recursive games is Everett's of the same title and is to be found in "Contributions to the Theory of Games, III." According to Everett, the game Γ has a value and it is to be found among one of the solutions of the equation

$$(1) \qquad x = \operatorname{val} \left\{ \begin{pmatrix} 1 & \lambda & x \\ \lambda & 1 & \lambda \\ x & \lambda & 1 \end{pmatrix} \right\}$$

where the right-hand side of the equation signifies the value of the game represented by the matrix enclosed in parentheses. If we let this matrix be A, then in order to solve (1) we are first faced with the problem of finding a function v of x and λ and probability vectors X and Y such that

(2) $$XA \geq (v,v,v) \quad \text{and} \quad AY^T \leq \left(\begin{smallmatrix}v\\v\\v\end{smallmatrix}\right).$$

v, X and Y, of course, are the value, a row optimal strategy and a column optimal strategy, respectively, of the game represented by the matrix A. It is easily verified that if $x \geq 2\lambda-1$, then

(3) $$v = \frac{1-2\lambda^2+x}{3-4\lambda+x}$$

(4) $$X = Y = \left(\frac{1-\lambda}{3-4\lambda+x}, \frac{1-2\lambda+x}{3-4\lambda+x}, \frac{1-\lambda}{3-4\lambda+x}\right)$$

satisfy (2); and if $x \leq 2\lambda-1$, then

(5) $$v = \lambda$$

(6) $$X = (0,1,0) \quad Y = (\mu,0,1-\mu)$$
$$\text{where } \mu \in \left[\frac{1-\lambda}{1-x}, \frac{\lambda-x}{1-x}\right]$$

satisfy (2).

Now let us return to equation (1). For $x \geq 2\lambda-1$ it becomes
$$x = \frac{1-2\lambda^2+x}{3-4\lambda+x}$$

and there are two possible solutions; namely,

(7) $$x = (\sqrt{2}-1) + (2-\sqrt{2})\lambda$$
and
(8) $$x = (-\sqrt{2}-1) + (2+\sqrt{2})\lambda .$$

Since
$$(-\sqrt{2}-1) + (2+\sqrt{2})\lambda = (2\lambda-1) - \sqrt{2}(1-\lambda)$$

we see that (8) only satisfies the restriction $x \geq 2\lambda-1$ when $\lambda = 1$, and in the latter case (8) can be included in (7). For $x \leq 2\lambda-1$ equation (1) becomes

$$x = \lambda$$

which combined with the restriction $x \leq 2\lambda-1$ becomes or

$$\lambda \leq 2\lambda - 1$$
$$\lambda \geq 1.$$

Hence the equation

$$x = \lambda$$

can only hold if $\lambda = 1$ and this case is also included in (7). Hence

$$x = (\sqrt{2}-1) + (2-\sqrt{2})\lambda$$

is therefore the only fixed point of the value transformation and thus is the value of the recursive game.

Also according to Everett, we know that optimal strategies for both players are to play the strategy

(9) $$(1 - \frac{\sqrt{2}}{2}, \sqrt{2}-1, 1 - \frac{\sqrt{2}}{2})$$

every time a new recursion comes up. We remark that if (7) is substituted in (4), then (9) results.

Now let us consider the game which is the same as before except that at the n-th recursion the game is truncated by putting zeros in positions (1,3) and (3,1). (Of course, the game may terminate naturally before the n-th recursion is reached.) If we define

$$x_0 = \begin{cases} \frac{1-2\lambda^2}{3-4\lambda} & 0 \leq \lambda \leq \frac{1}{2} \\ \lambda & \frac{1}{2} \leq \lambda \leq 1 \end{cases}$$

$$x_k = \frac{1-2\lambda^2+x_{k-1}}{3-4\lambda+x_{k-1}} \quad k = 1,\cdots,n$$

then the value of such a game is

$$v = x_n,$$

and optimal strategies for both players are to play (4) on the first n-1 recusions and play (6) on the n-th recursion. We indicate below what might be a proof of the above. It is by induction.

Case $n = 1$: Let us write

$$A_1 = \begin{pmatrix} 1 & \lambda & A_0 \\ \lambda & 1 & \lambda \\ A_0 & \lambda & 1 \end{pmatrix} \qquad A_0 = \begin{pmatrix} 1 & \lambda & 0 \\ \lambda & 1 & \lambda \\ 0 & \lambda & 1 \end{pmatrix}$$

The value of the game represented by A_0 is by (3) and (5)

$$x_0 = \begin{cases} \dfrac{1-2\lambda^2}{3-4\lambda} & 0 \le \lambda \le \dfrac{1}{2} \\[2mm] \lambda & \dfrac{1}{2} \le \lambda \le 1 \end{cases}$$

Replacing A_0 by x_0 in A_1 we get

$$A_1 = \begin{pmatrix} 1 & \lambda & x_0 \\ \lambda & 1 & \lambda \\ x_0 & \lambda & 1 \end{pmatrix}$$

We claim that $x_0 \ge 2\lambda-1$. In the case $1/2 \le \lambda \le 1$ where $x_0 = \lambda$ we have

$$1 \ge \lambda$$

$$\lambda+1 \ge \lambda+\lambda$$

$$x_0 = \lambda \ge 2\lambda-1.$$

In the case $0 \le \lambda \le \dfrac{1}{2}$ where $x_0 = \dfrac{1-2\lambda^2}{3-4\lambda}$ we have

$$(2\lambda-1)(3-4\lambda) \le 0 \le 1-2\lambda^2$$

since $0 \le \lambda \le \dfrac{1}{2}$. And, since $3-4\lambda > 0$ for these values of λ we have

$$x_0 = \dfrac{1-2\lambda^2}{3-4\lambda} \ge 2\lambda-1.$$

Hence $x_0 \ge 2\lambda-1$ and we have from (3) that the value of A_1 is

$$x_1 = \dfrac{1-2\lambda^2+x_0}{3-4\lambda+x_0}.$$

That the optimal strategies are as proposed follows from (4) and (6).

Case $n = k+1$ assuming case $n = k$ true: Let us write

$$A_{k+1} = \begin{pmatrix} 1 & \lambda & A_k \\ \lambda & 1 & \lambda \\ A_k & \lambda & 1 \end{pmatrix} \qquad A_k = \begin{pmatrix} 1 & \lambda & x_{k-1} \\ \lambda & 1 & \lambda \\ x_{k-1} & \lambda & 1 \end{pmatrix}$$

By our hypothesis we know that the value of the game represented by A_k is

5

$$x_k = \frac{1-2\lambda^2+x_{k-1}}{3-4\lambda+x_{k-1}}.$$

Hence from (3) and the restriction that is necessary in order for (3) to apply we can conclude that

(11)
$$x_{k-1} \geq 2\lambda-1 .$$

We claim that $x_k \geq 2\lambda-1$. If $\lambda = 1$ this follows from the fact that (10) reduces to

$$x_k = 1$$

and that

$$2\lambda-1 = 1 .$$

In the case where $\lambda < 1$ let us suppose on the contrary that

$$x_k < 2\lambda-1.$$

Then (10) becomes

(12)
$$2\lambda-1 > \frac{1-2\lambda^2+x_{k-1}}{3-4\lambda+x_{k-1}} .$$

Since $x_{k-1} \geq 2\lambda-1$,

$$3-4\lambda+x_{k-1} \geq 3-4\lambda+2\lambda-1 = 2(1-\lambda) > 0 \qquad (\lambda < 1).$$

Hence (12) can be successively rewritten

$$(3-4\lambda+x_{k-1})(2\lambda-1) > 1-2\lambda^2+x_{k-1}$$

$$(2\lambda-2)x_{k-1} > 4-10\lambda+6\lambda^2$$

$$x_{k-1} < \frac{2(1-\lambda)(2-3\lambda)}{2(\lambda-1)}$$

$$x_{k-1} < 3\lambda-2.$$

Since $3\lambda-2 = (2\lambda-1)-(1-\lambda) \leq 2\lambda-1$, we have $x_{k-1} < 2\lambda-1$ which contradicts (11). Hence $x_k \geq 2\lambda-1$ and by (3)

$$x_{k+1} = \frac{1-2\lambda^2+x_k}{3-4\lambda+x_k}$$

and, as in the case $n = 1$, the optimal strategies are as proposed from (4) and (6).

www.ingramcontent.com/pod-product-compliance
Lightning Source LLC
Chambersburg PA
CBHW081910170526
45167CB00007B/3226